PLANT SCIENCE, AGRICULTURE, AND FORESTRY IN AFRICA SOUTH OF THE SAHARA

PLANT SCIENCE, AGRICULTURE, AND FORESTRY IN AFRICA SOUTH OF THE SAHARA

With a Special Guide for Liberia and West Africa

By:

Cyril E. Broderick, Sr., Ph.D.
ASSOCIATE PROFESSOR

Delaware State University
Dover, DE 19901

Copyright © 2019 by Cyril E. Broderick, Sr., Ph.D.

ISBN:	Softcover	978-1-7960-7473-4
	eBook	978-1-7960-7472-7

All rights reserved. No part of this book may be reproduced or transmitted in any form or by any means, electronic or mechanical, including photocopying, recording, or by any information storage and retrieval system, without permission in writing from the copyright owner.

Any people depicted in stock imagery provided by Getty Images are models, and such images are being used for illustrative purposes only.
Certain stock imagery © Getty Images.

Print information available on the last page.

Rev. date: 11/26/2019

To order additional copies of this book, contact:
Xlibris
1-888-795-4274
www.Xlibris.com
Orders@Xlibris.com

CONTENTS

Foreword ix

Chapter I	Introduction 1	
Chapter II	Critical Considerations for Agriculture and Forestry in Liberia and West Africa 3	
Chapter III	Plant Science and the Structure of Plants 9	
Chapter IV	Soils and Soil Requirements Provided to Plants 15	
Chapter V	The Atmosphere in Which Plants Grow and Thrive 20	
Chapter VI	The Process of Photosynthesis 24	
Chapter VII	Respiration for Life 28	
Chapter VIII	Net Photosynthesis and Assimilation 29	
Chapter IX	The Productive Nature of Plant Agriculture 30	
Chapter X	Agricultural Plant Commodities and Products 31	
Chapter XI	Plants, Animals and Animal Husbandry 32	
Chapter XII	Animals and Animal Products for Profits 33	
Chapter XIII	The Nature of the Tropical Forest 34	
Chapter XIV	Forest Tree Species 35	
Chapter XV	The Forest Biome 36	
Chapter XVI	Pests in the Environment 37	
Chapter XVII	Sustaining an Environment for Productive Agriculture and Forestry 38	
Chapter XVIII	Development of Tropical Agriculture and Forestry 39	
Chapter XIX	Improvement of the Knowledge Base of the Population in Underdeveloped Countries 40	

Chapter XX	Biotechnological improvements of tropical plants	41
Chapter XXI	Establishment of food storage and food processing facilities	42
Chapter XXII	Silos	43
Chapter XXIII	Food Processing	44
Chapter XXIV	The Focus on Agriculture	45
Chapter XXV	Tropical Plant Species	47
References		61

To:

My children Cheryl, Cyril, Jr., Cynelsa, and Clyde; my mother, Sylvia; my late father, Nelson; my brother Jacob; and my sisters Vesta, Leona, Susan (late), Thelma, and Jackie

With Love,
Yours,
CEB

Foreword

This choice to write this book is recognition of the immense need to address the low level of productivity in agriculture, forestry, and the complementary professions that sustain our world. Africa is derelict in its delivery of the necessities for life, especially food, to its citizens – the people of Africa. There are too many reports of famine in African countries, and the need for food aid from developed nations to these countries make continuing reports of under-productivity and conditions of famine quite superfluous to the casual listener.

The neglect is now widely recognized, as reported by the Integrated Regional Information Networks (IRIN) in a report published electronically in a paper by the International Food Policy Research Institute (IFPRI), a US-based think-tank. The report says,

"African countries are not spending enough on agriculture and the overall productivity of the continent has dropped since the mid-1980s." I further states, "Since the 1960s, Africa has lost ground in the global marketplace. Its share of total world agricultural exports fell from 6 percent in the 1970s to 2 percent in 2007," quoting a paper entitled, Public Spending for Agriculture in Africa: Trends and Composition. Africa needs to do a lot better. West African countries have enormous resources that can provide for African citizens and enrich the people. A notable statement in their conclusion was, "Spending money on food production is critical in Africa, where 70 percent of people live in rural areas and depend on agriculture for food and income."

Developed nations, in contrast, are impressive by their commitment and their orchestration of family and corporation farms in the production of the required quantities of food and other products that are requisite economy. The production and processing of these commodities result in the desired products that secure additional income for these countries who utilize their surpluses for economic and political advantage. Surpluses from developed countries form major economic and political assets in their dealings with other nations and their people.

The prospect for a great future for West African agriculture is dependent upon the recognition of the contribution that the individual citizen, partnerships, and corporations can make in improving the production of essential commodities for the people of Africa, and with that recognition make deliberate, organized and systemized commitment to the production and processing sector. Farming is the first tier of productive investment, but only with good success, can people be released from the farm to engage in conservation and food processing and pursue other avenues for development. The perpetual dependence on others for basic food and agricultural supplies is the first prescription for economic disaster. Africa needs no such prescription.

Rice (*Oryza sativa*) is a primary staple of many countries in the tropics around the world, and it is also the primary staple in West Africa. The need to increase rice production is unquestioned; the demand for rice among Liberians, West Africans, and other Africans and almost all nations of the world, including China and the United States make the demand and utility of the crop very feasible, economically.

Cassava (*Manihot esculenta*) is a second important crop for Liberia and Many countries in Sub-Saharan Africa. Cassava is so very important that it is emerging to fill a larger role in West Africa and other countries as a crop the local population depends upon for food security. An additional positive factor about cassava is the number of food types that can be made from this root crop. The leaves are eaten as a vegetable, and the root is consumed as cooked cassava, and cooked cassava processed into fufu, dumboi, cassava fries, and the processed cassava farina.

Other root crops such as yams (*Dioscorea alata*) and sweet potatoes (*Ipomoea alata*) are also important, but their popularity, although clear, is significantly less than the relevance of rice and cassava. Other staples exist in tropical West Africa, and they include corn in Ghana, and eddoes (*Xanthosoma malfalfa*) in parts of Liberia and other African countries, including Nigeria and Ghana.

Crops grown to secure money for the purchase of other essentials are known as cash crops. Cash crops are very important, and they mainly non-food items and include sugar cane, rubber, coffee, cocoa, oil palm, and peanuts. Cash crops are important to the local trade or exchange process.

Animal production is complementary to plant production, and cattle, swine, goats, and sheep are also very important. These animals are supplied regularly to the local market, and their place, as well as the place of poultry cannot be underestimated.

Every allied occupation is relevant to this pursuit, and every individual with relevant knowledge, skills, and talents should recognize the commitment and become a working advocate in the pursuit. Productivity should be enhanced, and profits should be an intrinsic goal.

My recognition of the need for improved agricultural productivity has persisted over many years through undergraduate school at the University of Liberia, graduate schools at Iowa State University of Science and Technology, and the University of New Hampshire, both of the latter in the United States.. Additional stimulation has come through years of study, teaching, and research at the University of Liberia, the Firestone Plantation Company, Delaware State University, and most recently the William V. S. Tubman University. The necessity for this book, however, is the need for information to pursue agriculture by all who decide to accept the great challenge to pursue productivity in agriculture. That decision requires lots of knowledge of science in chemistry, physics, and biology, ability to carry out math calculations, and real knowledge of plants, animals, soils, light, and the above-ground atmosphere. This book provides a compendium of information for use by anyone who desires to know and practice any branch of agriculture or agriculture in general in meeting the demand for food crops and the commodities and products that result from such a demanding enterprise.

Good luck and best wishes go with you as you go through this book and your pursuit for productivity.

Chapter I

Introduction

The title of this book, *Plant Science, Agriculture and Forestry in Sub-Saharan Africa*, emphasizes a focus on a very critical facet of tenets that must form the focus in tropical Africa, and it is imperative that scientists, economists, sociologists, businessmen, and other professions address this most intrinsic component of the economic life and welfare of the people of Africa South of the Sahara. Famine with nutrient deficiencies among major portions of the populace has caused widespread nutritional problems among the people of Africa, and food scarcity continues to be the cause of famines and a scourge to the African population. Moreover, food production in tropical Africa must occur in an environment that considers the precarious nature of soil structure and texture as well as the tentative balance between incident daily sunlight, cloudy and rainy days among other physical components of the environment. The omnipresent competitive biological organisms, including different kinds of insects, bacteria, and fungi, as well as the roles of higher organisms that include birds, herbivores, rodents, other animals, and even man.

It is important that we recognize the indispensable nature of plants; therefore, one objective of the book project is to transmit the understanding of the scientific principles that govern plant growth and development. Plants in forestry, food crop agriculture, cotton and other commercial agricultural plants, as well as plants in rivers, seas, the oceans and other waterways would all be discusses. How plants function in different roles and different environments are major components of this text.

Animals with their production and management form a second focus. An additional objective is to ensure that people are the advocate for plants, especially in their production as feeds for animal consumption.

Wildlife is an important group of animals that serve in meat supply, where domesticated animals are inadequate.

Agriculture, forestry, and the environment. The designation to cultivate awareness and support the processes involved in plant science and the broad discipline of agriculture are intrinsic to this effort.

Chapter II

Critical Considerations for Agriculture and Forestry in Liberia and West Africa

<u>Plants</u>

Plants are the benchmark organisms on which agriculture must depend; therefore, our first focus is on plants. We will then proceed to discuss various other animals and their performance, and then follow with the different factors that affect agriculture and forestry in Liberia and West Africa. Moreover, plants are essentially the only group of organisms that would convert inorganic carbon from carbon dioxide of the air on water into sugars and other organic molecules that are synthesized in plant photosynthetic structures, the chloroplasts, and other cell, tissue and organ structures in plants.

Plants need an adequate supply of water from the soil, the essential mineral nutrient elements from the soil, water solution, or other cultivating medium in which the plant grows. Carbon dioxide is typically about 0.03% of the atmosphere, but its levels may rise to significantly higher levels when there is excessive evolution of CO_2 into the atmosphere. A few circumstances may also account for the loss of CO_2 from the air. A conducive environment in terms of air and soil temperature, the absence of contaminant gases in the air, an adequate supply of wavelengths of light that are required for photosynthesis, as well as an environment in which harmful biological organisms and agents are controlled and do not threaten growing and developing plants.

Lemmens et al. (2009) noted the important roles that timber species of tropical Africa can play in primary and secondary timber wood supply as well as use as medicinal plants, 'fuel plants,' and 'commodity groups,' including ornamentals and fruits.

Water

Water is a large component of plant tissue, and an adequate supply is needed to maintain a functional plant system. Jan Baptiste Van Helmont (ca. to 1644), one of the first experimenter with plants, wrongly concluded in the 17th Century that water was the principle of plant growth, not recognizing the total picture of what was occurring in the biochemistry and biophysics of plants during their development.

The rainy season provides water, and between 4000 and 5000 millimeters of rain may fall during the rainy season in coastal areas, while about 2000 millimeters, half the upper limits may fall in the areas of Liberia that are away from the coast. This precipitation rate is high, but it is concentrated in virtually six months of the year.

Plants uptake water mainly from the soil, but water is also obtained in relatively minor acquisition from moisture in the atmosphere.

Mineral Nutrients

Nitrogen is the most critically required nutrient element that is easily deficient in Liberian soils and in the Tropics. The atmosphere contains about 78 percent nitrogen, but that nitrogen is not available to most plants; legumes are one family of plants that have symbiotic relationships with certain bacteria that are able to pick up nitrogen from the air and convert the nitrogen into nitrates why they, the bacteria receive carbohydrates and other food compounds from the plant.

Other than nitrogen, the plant needs fairly large quantities of the elements phosphorus (P), potassium (K), calcium (Ca), iron (Fe), sulfur (S), and magnesium. It requires smaller quantities of the elements boron (B), chlorine (Cl), copper (Cu), manganese (Mn), molybdenum (Mo), and zinc (Zn). There are reports of special requirements for nickel (Ni) and silicon (Si) by certain plants, but the first 13 elements listed are the mineral nutrient elements that plants need for their growth and development. Three other elements, carbon (C), hydrogen (H) and oxygen (O) are obtained from carbon dioxide of the atmosphere and water from the soil and air that plants uptake, as stated above.

The Atmosphere and Carbon Dioxide

Carbon dioxide is the gas that is taken up from the atmosphere in photosynthesis, but the concentration of the gas in a typical atmosphere is low; therefore, it is important to choose plants based on the efficiency for the uptake and use of carbon dioxide by plants.

Selection of C-3, C-4 or CAM plants to take advantage of the metabolic concentration of carbon dioxide beyond the typical 300 ppm for the atmosphere is quite desirable.

The Soil

The soil is made up of mineral materials from parent rocks, and these mineral materials include a number of quality clays,. Clays are complex natural mineral structures of the soil that have the ability to hold mineral nutrients in the soil. The elements that are held by these clays are readily available to be picked up by plant roots that need the elements for the growth and development of the plant.

The soil also contains organic matter. Remnants of living materials or materials that were once alive are regarded as organic matter. These materials are rich in carbon and may have components of the elements hydrogen, oxygen, nitrogen, phosphorus, and sulfur, among other elements in the chemical and physical structure of its substance.

Decomposed organic mate4ials frequently become humus, complex organic molecules that are rich in many separated negative charges and some positive charges, making them ready bipolar and able to hold a rich supply of cations and anions on their structure. Organic matter is thereby able to hold many cations in place for uptake by plant roots, and it is now known that organic matter in soils raise the cation exchange capacity (CEC) of soils. Organic matter is also able to increase the water-holding capacity of soils, holding as much as 80 to 90 percent of its weight in water. Soils that hold a fair amount of organic matter are consequently significantly richer than soils with little or no organic matter.

Light

The requirement of light for plant growth and development was first presented by Ingenhousz and De Saussure, both of whom had become knowledgeable of the existence of the gas oxygen based on work by Joseph Priestley, an Englishman scientist, and Antoine Lavoisier, a French scientist. Priestley and Lavoisier had recognized that plants produced oxygen, which when breathe in by animals assured life to the animals. In experiments to confirm the results of Priestly and Lavoisier, Ingenhousz and De Saussure had found that the 'gas' that purified the air was not prod

Favorable Physical Environmental Conditions

The environment for agriculture in Liberia is tropical and is made basically of two seasons in contrast to the four seasons of the temperate environment. The two seasons of Liberia's tropical environment are the dry season and the rainy season. The four seasons of the temperate environment are winter, spring, summer and fall. Temperature, precipitation, light incidence and intensity, soil conditions are physical conditions that affect plants, animals, and the entire agriculture and forestry systems. We shall describe these conditions as this text progresses. With regard to temperature, it should be pointed out that the temperature range in the tropics is minimal and temperature maintains high above freezing and much lower than boiling. Freezing and boiling kill most species of the fungi, bacteria, and insects in their physically challenged phases in the life cycle.

Temperature range is generally above 10°C (50°F) and under 45°C (113°F). This 35 Centigrade-range (or 63 Fahrenheit degree range) is about all most tropical crop plants can tolerate. A few plant6s that have specialized mechanisms, such as the Crassulacean Acid Metabolism (CAM) plants may be able to withstand temperatures up to 50°C.

The high temperature is very significant in that bacterial activity and fungal activity are high. The resultant rate of decomposition of dead materials is so high that leaves and other organic materials decompose so fast that the benefits of humus and humic acid are very transient. The loss of organic materials by total decomposition is highest or lands that are uncovered or completely denuded. The result is that the land loses its ability to hold moisture and nutrient elements that are essential for plant growth and development. Nitrogen is a critical element for plant growth and development, and its loss under the circumstance is very high.

Protection from Biological Agents in the Environment

Plants grow from seeds or plant materials cut or explanted from other living plants. Because the contents of plants are carbohydrates, proteins, fats, and other important food substances, man is not the only organism that desires these good molecules. Cattle, goats, sheep, rabbits, mice and rats, insects, and bacteria and fungi are among the many groups of organisms that desire the contents of the plant. Plants need to be protected, consequently, form the many biological agents that have interest in them. This is a major challenge, especially when many acres or hectares or square kilometers of land have to be protected. Biological control uses one organism to feed on or prey on another organism (the pest). Legislation refers to laws passed to exclude recognized pests from an area. Mechanical control is effort exerted to intercept the pest; eradication refers to deliberate efforts pursued to completely get rid of the pests in question. Chemical control uses selectively toxic elements of compounds to reduce pest incident to a 'relatively safe' or innocuous level. The measures listed above are all means for the control or protection of plants.

Practical Responses to Critical Considerations

A. Build compost sites and use compost copiously
B. Have ready source of irrigation, whether from deep wells or from nearby streams
C. Keep a supply of fertilizers, natural as well as commercial chemical fertilizers
D. Keep the soil covered with trees or low-lying cover crops, whether they are vines or shrubs.
E. Encourage a free flow of air in the canopy of plants to discourage high moisture content or high humidity. High humidity is encouragement to the establishment of colonies of bacteria and fungi and egg deposits of insects.
F. Use forest species of strong economic value, including rapid-growing species, in providing soil berms and breaks against erosion as well as in providing shade for animals and in controlling various indicator conditions in the environment.

Chapter III

Plant Science and the Structure of Plants

Plants are fundamental to life, because they create the initial consortium of carbon molecules and all of the other complex compounds that are required by man and other organisms that feed on or parasitize plants. Plant science focuses on the understanding of the scientific phenomena that make plants function. There are physical, chemical, biological and complex other factors that affect plants. These factors may be centered in the soil complex, the above-ground atmosphere, water in its various forms and associations, and biological organisms, in all of the intricacies they present. Plants are biological organisms. They grow, reproduce, senesce, and die, and every facet of their lives must be studied and understood, because it is imperative that the essential productivity of these organisms are deciphered so that they can be harnessed as required.

Agriculture: Agriculture is an economic pursuit that utilizes scientific principles as the fundamental to the harnessing of resources in the production of plant and animal commodities that make products for human consumption, whether as food and beverages, clothing and textiles, or animal feeds and feeding stuff.

Forestry: Trees are the climax vegetation of much of the tropical landscape, and the canopies they form are shields that protect the land from the harsh environmental factors that are present in the tropical environment. These environmental factors include the high incident radiation, the high temperatures and drought, the periodic high levels of precipitation and the erosion they cause, and the diversity of wildlife they harbor.

The general populace of Liberia and most developing nations has very little professional knowledge of the forest, its trees, and its habitat. Other than the knowledge that trees make good firewood and that forest trees are valuable to the export market, most people cannot readily tell the difference between teak, mohagany, or ironwood trees. Moreover common names among the tree species can be very confusing. The ironwood tree in Delaware, USA, is referred to as the species *Carpinus betula* or *caroliniana, among other species.* whereas an ironwood tree in Monrovia, Liberia refers to the species *Lophira alata*. Their physical and cultural characteristics are completely different, and hence it is consequently important that efforts to improve the educational appreciation of forest species, forest management practices, and the identification and use of forest species among members of the general population can be a rewarding practice.

The Environment and Its Teetering Balance: Lessons from Temperate ventures suggest one option for dealing with many of these problems; however, too many of such adaptive ventures have led to mass failures. This text considers such mis-adventures and presents thought out and tried methods in providing solutions and directions to succeeding in the profitable production of trees and plant and animal commodities and products for the African and world populations.

The Science of Plants

Plants are a wonderful construct. Its leaves expose molecules of chlorophyll to radiation for absorption and use in the synthesis of sugars and other complex compounds in the functioning plant. It is consequently imperative that the plant scientist and the botanist understand the nature, use and adaptability of plants.

Biology: The biology of plants involves the exchange of gases, the utilization of water, salts, and elements in the reorganization of cells, tissues, and organs in functional organisms.

Chemistry: The elemental and molecular constitution of the cells, tissues and organs that make up the functional organism are based on certain phenomena that are controlled by the combining characteristics of elements, molecules, and entire arrays that create entities that conform to the requirements of their components.

Physics: The understanding of the physical attraction among entities of the environment is the focus of physics, and principles from the movement of water in plants to the absorption of solar radiation are questions that are answerable through their examination and study through the science of physics.

Mathematics: It is important that we can explain the relationships that exist among molecules, cells, organs, and individuals in the environment. Mathematics is that physical tool that can solve such inquiries. It is also quite clear that the measurement of these and other facets of plants and their life assure fuller understanding and greater insights into the processes that ensure the functionality of plants.

The Plant Organism

Plants are a very diverse group of organisms. They range from small traces of floating algae to tall voluminous evergreen trees that occupy planet Earth. In the more complex form, the plant has leaves, stems, and roots, that give rise to flowers, fruits, and seeds that grow, develop, and perpetuate the species.

Two complex processes are integral to the functional plant. It is important that we discuss these processes now, to enable us to appreciate the processes here to enable us to begin to understand the importance of the plant and its role in our world. The two fundamental processes are photosynthesis and respiration.

Photosynthesis

Photosynthesis is the biochemical reaction by which carbon dioxide combines with water to produce sugars. Oxygen is released as a by-product of the reaction. Carbon dioxide for the reaction of photosynthesis is naturally available in the atmosphere, where it occupies about 0.03 percent of the air. Water is present as humidity in the air, but the major significant source of water for the photosynthesis reaction is water that is absorbed from the soil through the roots of plants. Light is the energy source that propels the photosynthesis reaction. The sun naturally provides the energy, but artificial light can also provide the intensity of light that is required for photosynthesis. The outdoor environment is the typical system in which the photosynthesis reaction occurs, but artificial environments also host photosynthetic reactions. A summary of the photosynthesis equation is given below to show the reactants and the factors that are involved:

Light
$$6CO_2 + 6H_2O \xrightarrow{} C_6H_{12}O_6 + 6O_2$$
Chlorophyll
Carbon Water Glucose Oxygen dioxide

Respiration

The breakdown of sugars produces carbon dioxide and water, with the release energy for use by the organism. Respiration also produces mid-sized molecules that can be utilized in the construction or other molecules that are required in plant metabolism. The production of energy and the variety of other organic molecules for the plant is what is the major purpose of respiration. The summary chemical equation for respiration is given below.

$$C_6H_{12}O_6 + 6O_2 \longrightarrow 6CO_2 + 6H_2O + Energy$$ Glucose Oxygen Carbon Water dioxide

The general equation described above directly refers to aerobic respiration that results from specific activities of enzymes in the mitochondria. The final phase of respiration is referred to as oxidative respiration or oxidative phosphorylation, where water is produced by the protonation of oxygen that leads to the formation of water.

Energy is also produced in the process of fermentation whereby glucose is broken down to pyruvic acid which, if it does not react with Acetyl Co-A to enter the Kreb's cycle in the mitochondria, may be diverted to produce ethanol or lactic acid in a variation of the respiration pathway. This set of reactions that by-passes the Kreb's Cycle is known as anaerobic respiration. It produces energy also, but the quantity of oxygen that is produced is less than that produced through the Kreb's cycle reactions. As was explained to you above, it is clear that respiration is a process whereby glucose is utilized in metabolic processes, whether in the presence or absence of oxygen, and whether they occur inside or outside the mitochondrion.

Plants are that group of organisms that have the ability to carry out the process of integrating carbon from carbon dioxide into organic carbon for incorporation in the plant tissues. Other organisms including bacteria, fungi, snails, fishes, insects, birds, cattle, and man rely on plants for their ability to produce organic carbon molecules that are food and that are available for other uses by these and other organisms.

Some Promising Aspects of Tropical Plant Production

Currently, the most productive plant systems are in temperate regions, but tropical regions can also be very productive, particularly if genetic, physiological, and environmental factors are closely monitored and controlled for the production systems. Production systems in many countries in Asia and South America are doing quite well, and West Africa and all of tropical Africa can benefit and attain similar successes that have been obtained in those countries.

The Structure of Plants: Roots, stems, leaves, and fruits are the typical base structure of plants; however, it is the group of more developed plants (higher plants) that have all of the characteristics. Fruits generally provide for the sexual propagation of plants, because in fruits seeds are produced. The seeds produced are typically a product of the union of sex cells from a male pollen from the same plant or another plant and a female egg from the plant on which the fruit is borne. Plants also reproduce by non-sexual means. Plant cells, tissues, or cuttings can be made to multiply, grow and develop into wholly new and complete plants.

Yet, however, not all plants have reached the level that they can produce seeds. Some plants multiply by spores, while others only multiply by rudimentary sexual means of tissue separation. There are also the group of plants that are known as the Gymnosperms that bear seeds in cones. The structure of these and other plants will the discussed in this section of this text.

<u>Prokaryotic plants:</u> The karyote refers to the internal structure of the cell, and some cells do not have clearly defined, segmented or separated internal structures. Cells that have no clearly defined, segmented or separated internal structures are termed prokaryotes.

In this group are many species of plants, some of which live and survive as floating entities in moist environments.

<u>Ferns and Bryophytes</u>: Mosses, ferns and the non-vascular plants are organisms that are more structured in that they have leaves, stems, and roots, but their reproductive structures are spores not fruits and seeds. The pattern of growth and development of ferns and bryophytes will be discussed herein.

<u>Gymnosperms:</u> Gymnosperms are important species of plants, but they are typically very structured species, from the cells to the entire plant structure. They bear no fruits, but they bear cones in which their seeds are borne. Gymnosperm species account for thousands of trees and shrubs, many of which inhabit Temperate and Arctic lands.

<u>Angiosperms:</u> Angiosperms are flower-bearing species. They account for the majority of food plants, and they bear the nuts and seeds of all the fruit trees. Angiosperms are also the species that bear flowers that beautify gardens and lawns buildings and the home interiorscape.

Chapter IV

Soils and Soil Requirements Provided to Plants

Soils are the medium in which plants grow. The nature of any soil is dependent on the materials of which such soils are made. Such materials are referred to as the parent material, and parent materials vary from one region of the globe to the next. Below, we shall discuss differences in soils as well as the nature of the parent materials that make up soils in various countries in the Tropics and around the globe.

Miller and Gardiner (1998), like many soil scientists before them, classed the bulk of the soils of Liberia as Oxisols. The group Oxiosols is one of some 12 orders in the classification of soils. According to the text, soils in Oxisols order are often more than 3 meters (10 ft) in depth, well weathered (broken down by the environment) with only a few of the original mineral materials in their original form, and have low fertility. Soils in this order have a dominance of iron and aluminum oxide clays and are acid. The order Oxisols is a typical tropical soil, but there may be some smaller areas of Liberia that have soils that may be classed as Ultisols and Alfisols, but those areas are maybe a total of 10 percent of less than the total soil acreage of Liberia. Ultisols extend from eastern Guinea through Ivory Coast and Ghana. Eastern portions of Nigeria and northern portions of Cameroon are also classed in the order Ultisols. Areas north of the Guinea Plateau that run from Senegal and Gambia to northern Nigeria, Chad and the Central African Republic, and extending south into Togo and Benin are all grouped in the order Alfisols. It is notable that soils surrounding the Congo River in the center of Africa and the Amazon River in South America are largely Oxisols, like the bulk of the soils of Liberia, and both areas have a complementary presence of Ultisols. Ultisols and Alfisols are similar to Oxisols in certain characteristics, but soils in Ultisols group are strongly acid, and soils in the Alfisols group are slightly to moderately acid. You may check with your soil texts to obtain a more detailed description of soils of Liberia and West Africa.

The fact that soils of Liberia are low in fertility, acid, but relatively deep is a mixed prognosis, but with adequate thought and management, there is significant hope in enhancing the soil, raising its fertility, and making it wholesome and nutritious for agricultural and forestry species.

Improvement of tropical soils involves the provision and incorporation of organic materials which would serve to enhance water holding capacity, base elements incorporation, and promote cation exchange capacity. The acquisition and use of inexpensive sources of fertilizer and commercially available fertilizers, especially nitrogen fertilizers, is a significant enterprise that must be pursued to develop soils that would be nutritious to plants.

Furthermore, organic materials must be preserved. Uncollected and unpreserved organic materials, especially when in hot and unprotected environments are easily decomposed into the tropical air, releasing important volatile organic compounds from dead plant and animal residues on the soil in top layers of the soil into the atmosphere. These materials, when made gaseous, become unavailable to seedlings and plants that are growing in the soil. Crispeels and Martin (2000) presented a table that describes the relative rate of decomposition of organic materials in different ecosystems around the globe. They report that although tropical rainforests produce some 14,000 kilograms of leaves per hectare compared to temperate deciduous forests and temperate conifer (evergreen) forests that produce only 4,500 and 2,700 kilograms of leaves per hectare, respectively, it takes only 1.7 years for the organic matter that was produced to decay in the tropics, whereas deciduous temperate organic litter took 4 years and conifer forest litter was lost in 14 years. It is consequently obvious why litter accumulates in temperate coniferous forests while it disappears quickly in the tropics.

The area between the latitudes of the Tropic of Cancer (23° 27' N) in the Northern Hemisphere and the Tropic of Capricorn (23° 27' S) in the Southern Hemisphere essentially categorizes the Tropics. In actual terms, the region of the Tropics (tropical climate) stretches between 15°N latitude and 15°S latitude. Much of the terra ferma of Africa, Asia, Central and South America, and parts of the northern portion of the continent of Australia are included in the tropics.

The origin of tropical soils is a little variable, but these origins are classified appropriately in texts on soils. The nature and properties of tropical soils depend on their origin, their composition, and their environment. A critical consideration in the use of tropical soils in agriculture, however, is recognition of the integral components of the soil, including its content of organic matter.

The fact is that organic matter decomposition in tropical soils is rapid, and that the resultant lack of organic matter affects the availability of water and nutrients. to plants is often sub-optimal, causing a lower productivity of plants grown under such conditions. This concept is presented in a table in Chrispeels and Sadava's text, *Plants Genes and Agriculture*, in which they presented comparative data from the Tropics (tropical rainforest), the Temperate (deciduous forest), the northern coniferous forest, and the Arctic (tundra). For the same kilogram of organic matter, it would require 50 years for decay in the Tundra, 14 years in the northern conifer forest, 4 years in the temperate deciduous forest, and only 1.7 years in the Tropics. This means that the Tropics would lack this essential soil component for the growth of crops.

Major natural differences between tropical and temperate agriculture are based on the nature of the parent materials of the two soil groups. The performance of agriculture in the temperate and the tropics is based on the natural difference in parent materials of temperate soils versus tropical soils as well as variation in the natural accumulation of organic materials between soils in the two environments. The environmental conditions of the tropics facilitate rapid decomposition, and tropical soils are consequently especially low in organic matter.

The low organic matter levels are exacerbated when the land is bulldozed, totally cleared, and then plowed under, without supplementing the soil with new organic materials and without the establishment of soil cover before windstorms and the rain. There must be recognition of the need to protect the soil by the addition of organic matter to the soil as well as to cover the land with cover crops or trees. This is most important in preserving the land and for the improvement of agriculture and agricultural production in the tropics.

Other factors that affect soil constitution include the constitution of the original parent materials that formed the soil as well environmental factors that have affected the formation of such soils. A first environmental consideration in the tropics is precipitation as rainfall. With rainfall, there is severe erosion and leaching with the loss of soluble nutrient elements, calcium being one of the major nutrient elements that is eroded, leached out, and lost from tropical soils. In contrast, most temperate soils have precipitation as rainfall, snow, and ice, and there is less of the physical dynamics that deal with erosion and leaching. Hence, temperate nutrients remain in place in the soil, while tropical soil nutrients get lost from the soil. Calcium is one element that is widely lost from tropical soils; therefore, in addition to the lack of organic materials, tropical soils need the addition of calcium, especially, to the topsoil.

For tropical soils, the question is how do we remedy the other soil problems? Because many of these problems can be managed and solved by careful planning, it is important that extensive management practices are incorporated in every farming venture. There should be regular measurement of measure soil nutrient content, soil pH, and to contrast such content with the needs of the plant species that would be planted and established on the tropical soil in question.

The nature of parent materials, the amount and frequency of precipitation, temperatures and their range, and other climatic factors must also be measured and monitored. Other biotic influences, including the presence of various soil organisms, land plants, and animals on the soil affect nutrient status and their availability for uptake.

Remedies for deficiencies in tropical soils include the purchase and provision of chemical fertilizers, as well as long term rehabilitation, including the use of fallow, to replenish soil nutrients that have been lost over years of mis-management of improper control of the soil resource.

Chapter V

The Atmosphere in Which Plants Grow and Thrive

The shoots of plants live in an above-ground atmosphere and plan roots live in a below-ground atmosphere. The portion of the plant we typically see, however, is that greater portion that is above the soil. We will therefore discuss the above-ground atmosphere firstly, and then we will discuss the below ground atmosphere in characterizing the environment in which plants grow and develop.

<u>The Above-Ground Environment:</u>: The above-ground environment is the atmosphere in which plants and the human population as well as most other organisms reside. It is located on the Earth's crust. The above-ground atmosphere is gaseous, and its average condition is that it is composed of 78 percent nitrogen, 21 percent oxygen, about 0.7 percent argon, and 0.03 percent carbon dioxide, a most important gas in the chemistry of life, and a variety of other components. Taiz and Zeiger (2002) reported that atmospheric carbon dioxide had risen to 370 ppm and that CO_2 levels in the air was rising at about 1 ppm per year, mainly due to the burning of fossil fuel.

Water vapor is a component of the atmosphere, and its content of the air is measured in terms of the relative humidity of the air. The relative humidity is the proportion of the air that is occupied by gaseous water in comparison to the maximum amount of that gaseous moisture that can be held in the air at that temperature and pressure. We will discuss the intricacies of the nature of the above-ground atmosphere as this topic reverberates in other parts of this text.

The Below-ground Environment: The below-ground environment is typically composed of the soil matrix, water, soil-trapped gases and a multiplex of bacteria, viruses, earthworms, ants and other animals, and materials from formerly living organisms (also known as organic litter), among other things. Plants have adapted to live and thrive in a variety of such environments; yet, the characterization of each rhizosphere is very important to understanding the performance of various plant species.

In the previous chapter, we discussed the physical component of the soil, including the mineral matter and organic matter. Emphasis was not put on the pore space of the soil. The pore space varies as to the relative soil components as to the proportion of sand, silt and clay, as described by the soil texture. There are other soil characteristics measured in terms of the soil bulk density or the soil particle density.

Pore space of a soil affects the bulk density of the soil, and that pore space is relative to the amount of air and water that the soil may hold. Plant roots and other organisms that live in the soil need air to survive, and roots and these organisms produce carbon dioxide and other gases that must be released from the soil to keep the soil habitable and supportive of living roots and other life. Carbon dioxide (CO_2) at 3 to 5 percent atmospheric concentration has a direct inhibitory rate on respiration (Taiz and Zeiger, 2002); therefore, it is not to anyone's advantage to have the level of carbon dioxide rise in the soil. Oxygen in the soil and root zone is very important to support respiration, an essential process to plant and animal life.

The effects of roots and plant and animal life on soils is study in the specialty known as edaphology, a branch of soil science in which soil is studied as a habitat for organisms and a medium in which plants grow (Taiz and Zeiger, 1998). The other branch of soil science is pedology, where soil is studied in terms of its geologic entity, including its origin, morphology, geography, as affected by geography, climate, living matter, parent material, relief, and time, among other factors.

Rainfall, temperature, relative humidity, air borne pests, wind, and several other factors are relevant to this chapter, but they will be addressed in other portions of this text.

The Above-Ground Environment of West Africa

West Africa is tropical, with temperatures that range from above freezing, around 40 to 50 °F to highs of well over 100 °F. Schulze (1973) reported that whereas cities along the coast of Liberia get an average of nearly two hundred inches of rainfall in a year, Abidjan gets only 81 inches, Lagos 72 inches, and Accra only 28 inches in the year. He credits the differences in rainfall obtained to the orientation of the south-west monsoon winds that interact between the Atlantic Ocean and the land mass of West Africa.

The Above-Ground Environment of Liberia

The rainy season and the dry season are the two seasons of the climate of Liberia. Generally, it is stated that the rainy season runs from May to October, and the dry season runs from November through April. Temperature levels and temperature changes during the seasons are two other factors in the above-ground environment that are important. High and low temperatures and daily and monthly temperature ranges affect the activities of various organisms, from bacteria and fungi, to insects and mammals in their survival patterns.

Precipitation: Despite this fixed categorization, there is some variation in the beginning of the rainy season and its end, whether beginning in April or May or ending in October or November. Maps of rainfall in Liberia, as presented by Shulze (1973) show that average rainfall in most of Liberia is between 2000 mm (78.7 inches) and over 4500 mm (177.2 inches) per year. Coastal Liberia has the wettest climate, but even precipitation varies from the western border to the eastern border of Liberia. The average annual rainfall for Robertsport and Monrovia are the highest, reaching rates around 5200 mm (205 inches) of rain per year.

<u>Temperature</u>: Average monthly temperature ranges between 20 °C (68 °F) and 30 °C (86 °F). The average indicates that temperatures rise above and below these temperatures, but those mean values give good indications about how much heat or how much cold that plants and animals that live in the region need to tolerate.

<u>Humidity</u>: The amount of moisture in the air is humidity, and relative humidity is the percentage of the amount of moisture in the air compared to the maximum amount of moisture that that air can hold at a specific temperature. The value of the relative humidity is very important, because the higher the relative humidity, the easier it is for condensation to occur, thereby providing moisture for the growth and development of fungi, bacteria, and other pests that infect plants and animals.

Relative humidity values are significantly lower during the dry season than during the rainy season. That means that infection of plant and animal tissues is less likely during the dry season than during the rainy or wet season.

Chapter VI

The Process of Photosynthesis

Photosynthesis is that biochemical and biophysical process whereby light energy is transformed to biochemical and biophysical energy for use by the plant. Plants use energy and products of the process in establishing their productivity and physical structure. Man, animals, and various pests also claim products from plants for use in their growth and development. The products of plants and their essence to all living matter make the understanding of plant production systems extremely important and essential.

This focus on the process of photosynthesis is deliberate, because the subject is rarely emphasized despite its importance and significance. Although our coverage here is limited, we shall be emphasizing features that should be recognized in our thinking and discussion of this important topic.

Photosynthesis takes place in plant leaves and other plant materials that contain chlorophyll and relevant complements for the process to occur. Because the leaf is the typical structure in which photosynthesis occurs, let us examine the architecture and construction of the plant leaf.

The Plant Leaf: Composed of palisade parenchyma and spongy parenchyma in a mesophyll between the abaxial and adaxial epidermal tissues, the plant leaf is the center for the synthesis of carbohydrates from carbon dioxide of the atmosphere and water that was uptaken through the plant roots. The leaf has stomata that admit air containing carbon dioxide. The stomata also release oxygen, a product of the photosynthesis process. In the absence of light, when photosynthesis does not take place in most plants, larger quantities of carbon dioxide are expelled as a result of respiration and the recycling of carbon dioxide that could not be fixed in the photosynthetic process.

Plants have different adaptations to photosynthesis, and there are three different pathways for photosynthesis that have been characterized in plants. They are the Calvin cycle photosynthesis (C-3 photosynthesis), the Hatch-Slack Pathway (C-4 photosynthesis) and the Crassalucean Acid Metabolism (CAM) pathway. Each pathway shall be explained briefly below to distinguish the various ways in which carbon dioxide is processed in photosynthesis.

C3 Photosynthesis: The Calvin Cycle is the basic photosynthesis process that occurs in plants. In this process, a primary event is the reaction of adenosine triphosphate (ATP), a molecule formed in the light reaction of photosynthesis, with the sugar phosphate compound, ribulose-5-phosphate, to form ribulose-1,5-biphosphate. Adenosine diphophaspate (ADP) is a product of the reaction. The molecule ribulose-1,5-biphosphate (RUBP) is a major product of the reaction. The next critical reaction is when the molecule ribulose-1,5-biphosphate reacts, under the catalytic influence of the enzyme ribullose-1,5-biphosphate carboxylase/oxygenase (RuBPc/o), with carbon dioxide (CO_2) to form two molecules of 3-phosphoglyceraldehyde (3-PGA). In the experiments, radioactive carbon dioxide ($^{14}CO_2$) was used. The enzyme RuBP c/o is a carboxylase because the reaction can take place with carbon dioxide as a reactant or an oxygenase, because the reaction may also take place with oxygen as a reactant. When carbon dioxide is the reactant with ribulose-1,5-biphosphate, the products are two molecules of 3-phosphoglyceric acid (3-PGA). This is a primary reaction in photosynthesis, and it forms the basis of what is termed C3 photosynthesis. It is termed C3 photosynthesis, because the first isolated product from the reaction is the C3 compound 3-phosphoglyceric acid (3-PGA). This basic process of C-3 photosynthesis is what was described by Calvin and co-workers, and it is the discovery for which Dr. Melvin Calvin was presented the Nobel Prize for Chemistry in 1961. Figure XXYZ provides a summary diagram to show the initial reactions that have been described here. The full Calvin Cycle would be presented later in this chapter in Figure XXYZ123. It should be pointed out here that elucidation of the pathway of carbon in photosynthesis was possible because of the discovery and use of radioactive carbon, carbon-14 in biochemical chromatographic analysis of the products of the photosynthetic process.

C4 Photosynthesis: The discovery of the pathway of carbon in photosynthesis had taken philosophers and scientists through many hypotheses and theoretical frameworks, and many scientists were satisfied with the results but ready to test the accuracy of the new discovery. Dr. H. P. Kor4tschack and colleagues, working with sugarcane in Hawaii, found that the first product of the reaction was a c-4 compound, not a c-3 compound, as had been found by Calvin and co-workers. Soon thereafter, other colleagues found similar discrepancies with corn (maize) and certain other plants. After exhaustive study, Dr. M. D. Hatch and Dr. C. R. Slack of Australia were able to confirm that the first product upon the admission of carbon dioxide into the leaf of sugarcane, corn, and a select number of species was a C-4 compound. Their work revealed that the C-3 carbon cycle exists in these species, but such plants had an extra mechanism for the accumulation of carbon. This first step in the accumulation of carbon is facilitated by the enzyme phosphoenol pyruvate carboxylase (PEP_c) in the synthesis of the C-4 acid compounds malate, oxaloacetate or aspartate. The essence is that plants that carry out this mechanism are able to decarboxylate the four-carbon compounds to release carbon dioxide in the bundle sheath were the enzyme ribulose-1,5-biphosphate carboxylase/oxygenase is able to pick up the released carbon dioxide in the synthesis of the C-3 compound, 3-Phosphoglyceric acid. The design of C-4 plants to anatomically have a bundle sheath that is separate from the rest of the mesophyll in the leaf is a major adaptation of C-4 plants. In summary this is the C-4 Process: Carbon dioxide that was uptaken through the stomata of the leaf were fixed into C-4 acids by the phosphonelpyruvate carboxylase enzyme. The C-4 acids are able to transverse the walls of the bundle sheath, where upon entry, released the CO_2 molecule from their structure. The CO_2 then reacted with RuBP molecule to form the typical 3-PGA molecule, as seen in C-3 plant photosynthesis. The structure of C-4 plant leaves is seen in Figure XXYZ1000A, and the schematic of the pathway is given in Figure XYZ1000B.

CAM Photosynthesis: A second modification of the Calvin cycle was found in plants of the Crassulaceae, a family of succulent plants. The genera Crassula, Kalanchoe, and Sedum are all members of this family that lends lends its name to this second modification of the photosynthetic pathway, namely the Crassulacean Acid Metabolism (CAM). Plants that exhibit this pathway typically close their stomata during daylight hours and open their stomata at night. This is the complete opposite stomata disposition in comparison to C-3 plants.

When the biochemistry was investigated, it was soon discovered that when the stomata were open at night, plants accumulated malic acid from the reaction of carbon dioxide with phosphoenol pyruvic acid, as facilitated by the enzyme phosphoenolpuryvate carboxylase. This enzyme is only a carboxylase and has no oxygenase activity; therefore, it does not react with oxygen that is more abundant (21%) of the atmosphere, when compared with the concentration of carbon dioxide (between 0.035 and 0.04 %) in the atmosphere. In the case of CAM plants, there is temporal (day versus night) biochemistry. During the day, the malate breaks down, releasing a molecule of carbon dioxide to react with RUBP, under the influence of the RUBPc/o enzyme in the production of 3-phosphoglyceric acid. The CAM pathway is presented in Figure XXYZ2000

Chapter VII

Respiration for Life

Respiration is that biochemical and biophysical process that enables energy to be produced for internal use of the organism. Typically, biomolecules react and are transformed with the release of energy and chemical moieties that serve structurally or functionally in the sustenance of the individual. Final breakdown products of respiration include carbon dioxide and water, and simple equation of the breakdown of glucose through its combination with oxygen to yield water and carbon dioxide, with the release of energy is the typical summary equation of respiration. Reactions that facilitate respiration are intrinsically important, and the importance of respiratory activities in cells and organisms cannot be underestimated.

Respiration provides energy to the cell for the many processes that occur in the cell.

The process of respiration produces the backbone of many molecules that form the structural basis of too many compounds that are used in the biochemistry of the plant.

Chapter VIII

Net Photosynthesis and Assimilation

When total photosynthesis exceeds total respiration, the net difference is positive and this is net photosynthesis, a mass of materials that become a physical addition to the body of the organism. This is the assimilate, and assimilation is a very important component of growth.

Growth is that irreversible increase in size, and the rate of growth depends on the levels of assimilation that are attained by organisms. Efficiency in photosynthesis and a reduction in total respiration favors assimilation. This will be discussed in further detail later in this text.

Chapter IX

The Productive Nature of Plant Agriculture

Plants are a transition organism, because they are able to transform inorganic carbon dioxide into organic molecules for use by the plant and all other organisms that consume plants directly or indirectly. This is the basis for the productive nature of plant agriculture. All of the vegetables, agronomic crops, as well as fruits and specialty crops, owe their productivity to this ability of plants to photosynthesize and produce new organic materials.

Chapter X

Agricultural Plant Commodities and Products

Commodities refer to categories of agricultural produce that have not been processed on taken through any value-added improvement. Cocoa beans, coffee beans, rice, rubber latex or coagulates, and even bags of cassava, beans, or baskets of tomatoes are all commodities.

Products refer to categories of materials that have been improved by manufacture or other value-added processes. Commodities are priced and sold at very low base prices, but products are sold at value added prices, and such prices may have little or no reflection on the purchase prices of the commodities.

Chapter XI

Plants, Animals and Animal Husbandry

Plants provide the base feedstock for animals, and the production of animals have their basics in the provision of animal feeds and feeding stuffs directly from plants.

Chickens and eggs in poultry husbandry form a large sector of animal husbandry. Pigs raised in piggeries, cattle (bulls and cows) in open range and in barns, and even goats, sheep, rabbits, and specialty breeds such as ducks, turkeys, geese, ostrich, and buffalo, as well as fish-farming are important agricultural investments that depend on plants.

Because all of these animals rely directly or indirectly on plants, the provision of animal feed is very important to the success of animal husbandry enterprises. Overly high cost of animal feed can easily make investments in raising pigs, poultry birds or cattle less attractive to the farmer or investor.

Chapter XII

Animals and Animal Products for Profits

Eggs are major money-makers for farmers, and so are meat from pigs or cattle, for example. There is significant difference in income from processed pig meat that produced sausages and bacon than the pork from freshly-slaughtered animals.

Production and processing provide excellent opportunities for high profits.

Chapter XIII

The Nature of the Tropical Forest

The climax vegetation of Liberia is tropical rain forest. The nature of the physical environment that include the soil, biological entities that occupy the land, and traditional land use practices that sustain that landscape over the years need close monitoring. The use of large and complicated machines is bound to complicate the balance of the land and challenge the different factors as they seek to work in harmony.

That scenario that recognizes the threats in the use of large tractors and chemical and biological entities that readily transform tens and hundreds of acres of land at a time into new ecosystems. The forest biome that kept has for many years keep an ecological balance is now very threatened by deforestation and denudation of the land, because of the thoroughness of the removal of ground cover and topsoil.

Chapter XIV

Forest Tree Species

Forest tree species serve as materials for construction of homes, buildings, scaffolding, furniture, and a multiplicity of other uses. Te quality of wood products is variable and many tropical species have excellent properties and are widely demanded. These resources are also widely sold as commodities, but their potential as excellent forest products cannot be underestimated. Timber and plywood are two readily manufactured products from the forests of Liberia.

Chapter XV

The Forest Biome

The Sapo National Forest is one of the largest forest tracts that remain in tact in Liberia, but there is serious incursion, and it must be protected. Similarly, large tracts of forests have been exploited in various parts of Liberia. It is important to totally assess the forests of Liberia, and confirm plans for their strategic management over the coming decades, at least.

Chapter XVI

Pests in the Environment

Insects are a majority group of pests in the Liberian ecosystem, and there are lists hundreds of insect species that occupy the soil, the air, and various organisms. Bacteria, fungi, nematodes, and even larger organisms may be pests of agricultural and forestry species. We will focus on these pests generally, and then we will point out some of the pests that should be monitored closely in the efforts to improve agriculture and forestry in West Africa.

Chapter XVII

Sustaining an Environment for Productive Agriculture and Forestry

Arable land is soil that can produce a crop, and this means that the soil has the ingredients that would support plant growth and development. Such land has an adequate supply of nutrients, a soil of water and the constituency and ability to hold moisture to ensure that the plants thereon can secure its demand for moisture. Non-physical factors are also important, and temperature is one such factor. A very hot environment would evaporate the supply of water so very quickly that the plant would be unable to secure and maintain an adequate supply of water to assure the survival of the plants being cultivated. Too cold an environment would make it impossible for available water to be transported to portions of the plants where water is needed to complete some metabolic or physiological process.

Soils of arable land have a fair content of organic matter in its make-up, and it is important that organic soil content be maintained.

Chapter XVIII

Development of Tropical Agriculture and Forestry

Tropical Agriculture has to develop rapidly, because as the world population grows, food and other resources for superior quality of life will become increasingly scarce, more expensive, and very unavailable in the Tropics. Moreover, the resources of agriculture in the tropics intrinsically include lands that are in forest, lands that have to be included in the planning of agriculture and which are necessary in developing sustainable agriculture in the tropics. Several reasons for the development of agriculture and forestry are clearly evident. They include: the currently low existing technical knowledge per capita in most tropical countries, the gross lack of manufacturing industries, and the exorbitant negative balance of trade for food crops and basic other resources, including energy sources, for almost all tropical countries.

The need is for accessibility to food supply, and this requirement is crucial; otherwise, there will be huge catastrophes in the loss of life among millions in the underdeveloped world. The question is how can we avert this pending crisis with the development of tropical agriculture?

Chapter XIX

Improvement of the Knowledge Base of the Population in Underdeveloped Countries

The first route is through the improvement of knowledge of agriculture and other live-based sciences among the general population. This can be done by introducing general science and agriculture in the early school years as well as in the general education requirements in college programs.

Chapter XX

Biotechnological improvements of tropical plants

The improvement of crops by genetic engineering techniques is leading the way to a new realm of hope for food and agriculture in the developing world of the tropics. Today, many tropical species are important world crops, and the indication is that opportunities to improve those species currently exist. Such crops inc;lude the coffee (*Coffea arabica* L., *Coffea robusta* L., *Coffee liberica*, L. and others), Cocoa (*Theobroma cocoa*), Banana (*Musa paradisica* and *Musa cavadishii*), Coconut (*Cocos nucifera*), Oil palm (*Elaeis guineensis*), Bitter leaf (*Vernonia amyglina*), Cowpeas (*Vigna unquilculata*), Cassava (*Manihot esculenta*), and many other crops are amenable to improvement through genetic engineering.

Chapter XXI

Establishment of food storage and food processing facilities

Food preservation is an essence for agriculture in the tropics. The harvest is the culmination of a tremendously energy-intensive project. There was land preparation, that included some plowing and harrowing, sowing and planting, fertilization, provision of adequate exposure of the plants to sunlight, good air, and a favorable environment, including the control of bacterial, fungal, viral, and insect pests.

Only when surpluses of the harvest can be stored and preserved would there be some food security; hence, there is urgent need for policy and serious efforts to establish some level of silo storage and food processing.

Chapter XXII

Silos

Silos are everywhere in US agriculture, but they are scarce or virtually absent in Liberia and too many tropical countries. The modern silo is complicated and controlled by computers, but the technology is varied, and there can be silos of various levels of sophistication.

When surpluses are saved, there is the opportunity for food security and the confidence among the population that they can delve into other endeavors without a worry about the availability of the next meal. This is in contrast as to whether a ship will dock with rice, flour, onions, tomato or beans, or whether the price of the different foodstuffs would rise quickly because of scarcity.

Chapter XXIII

Food Processing

At harvest, the plant, its fruit or other product is at its optimal attractiveness; soon thereafter, it naturally begins to lose its luster. Additionally, the nutrient content of such products are desired by insects, bacteria, fungi, and other organisms, including rodents; hence, such products may rot, dehydrate, and go through various degrees of deterioration. To counteract this, these products must be dried or dehydrated, fried or stored in oil, cooked, pasteurized, salted, pickled or stored in vinegar, or carried through various processes to maintain their edibility and nutritional value.

Chapter XXIV

The Focus on Agriculture

Agriculture will continue to form a major sector of the Liberian economy in the foreseeable future. This reality is clear when all of the indicators of development show that Liberia ranks in the lowest tier of development, the most telling being the literacy rate. Literacy correlates well with per capita income, health care levels, and life expectancy, among other things; therefore, it is clear that Liberia is not poised to receive the levels of benefits that current world technological innovations present. Current technological advances include computers and information systems, biotechnology, chemistry, medicines, mechanical and automotive design and manufacture, mechanization of agricultural processes, and various other industrial processes.

The reality is that the agricultural system is widely rudimentary, where cutlass agriculture is the norm. Furthermore, slash and burn systems around the tropics eliminate important forest resources and expose soils to rapid degradation. Moreover, the self-sufficient farmer or farm family is poor, and the very low availability of cash ensures barter trade with a rare purchase of goods and services that require a large capital outlay. With this reality is the fact that there is very minimum use of fertilizers, pesticides, and machinery. Conservation practices to secure harvest are quite poor, and modern silos are rare to non-existent in many tropical countries.

Because of the foregoing, it is imperative that efforts are directed to support the development of agriculture, firstly by the use of requisite knowledge and ingenuity that are available to build the larger arena around which strong economic activities develop around agriculture.

Chapter XXV

Tropical Plant Species

<u>The Natural Tropical Forest</u>: A natural vegetative cover for much of the tropics is forest. The forests are typically hardwood species that grow to form canopies that are tiered in at tens to dozens of feet in height. These canopies cover the soil and protect it form the intense solar radiation and tropical heat that reach the earth.

The canopy also intercepts incident rainfall, protecting covered soil from the hard raindrops of incessant precipitation during the rainy season. This is effective against excessive leaching and harsh disturbance of the O level and the topsoil.

Forests of the tropics produce some very important species for local use as well as for the world market. Table _____ lists some of the species that are important in the world market, as well as some species that are valued in local tropical markets.

Forestry species are valued firstly for the logs that they produce. The logs are sawed into timber, and some of the timber products are made into plywood and other processed products.

The value of trees of the forests is greater than merely the cash value in the logs and timbers. They play an indispensable role in land resource conservation. Deforested and denuded lands lose their value to forestry and agriculture quickly, and it is very expensive to virtually impossible to reclaim land that is overexploited and degraded.

Agricultural Plant Species

Dominant tropical species: Several plant species are major crops in tropical agriculture. They feed the population, and they provide animal feed and feeding stuffs for the animals.

Rice (*Oryza sativa*) is the most prominent cereal grain in the Tropics. It provides the largest proportion of the daily caloric intake of people who live in this part of the world. Other cereal crops include maize or corn, millet, sorghum, and a few other species.

Cassava (*Manihot esculenta* Cranz) is a second important and predominant species. Its root is the major food commodity from which many foods are prepared. The cassava leaf is also used widely for food.

The Yam is Dioscorea alata, a species that is grown for its underground stem. The International Institute of Tropical Agriculture (IITA) in Ibadan, Nigeria, has done a lot of work on the improvement of this important West African crop.

The Eddoe is *Xanthosoma malfalfa* and *Colocasia esculenta*, and this plant species is also grown for its underground stems that are valued by diverse groups of consumers.

Legumes: Tropical legumes of major significance are the pulses, peanuts, and a few other species of the Leguminosae. Important food and feed legumes range from the local countrypeas, pigeonpeas and cowpeas (*Vigna unquiculata*) to the exotic soybean (*Glycine max* L.)

Primary Food Crop Plants: Rice (Oryza sativa) is a staple or major food of many Asian and African tropical countries, including Liberia, Sierra Leone, Thailand, Nigeria, and the Philippines.

Primary Cash Crops: Certain plants are grown for sale and are referred to as 'cash crops.' The primary cash crops are grown, harvested, and semi-processed for the export market. The most widely recognized cash crops are coffee, cocoa, piassava, palm oil, palm kernel and palm kernel oil, coconut and coconut oil, and a variety of spices, among other crops.

Locally Consumed Cash Crops: Some cash crops are locally consumed, but their marketing is more direct. They include many of the spices, greens and other vegetables, sugar cane, and many local produce that are in demand for local consumption.

<u>Sugar cane:</u> Sugar cane (*Saccharum offinarium*) is a very important crop species in the tropics. In Cuba, it is a major crop that is the base of the sugar industry. In Hawaii, the crop is produced in one of the most efficient agricultural programs. In almost all poor tropical countries, the sugar cane is also a snack food that is relished by children and adults. Sugar cane is also the major raw material in the production of alcohol for drinking and for gasohol, where it is produced in the tropics.

<u>Peanuts</u>: Peanuts (*Arachis hypogeae*) are a legume crop, and it is produced through the length and breadth of the tropics, where it is consumed in a variety of preparations. It is popular as a roasted or parched snack that is sold at roadside stands and by individual vendors.

The peanut is a rich although unusual plant. The nut is a fruit, but after the plant flowers, and upon fertilization of the flower, the ovary is deposited underground, by a very unique process, where the peanut grows and develops to maturity.

Exported Cash Crops

Rubber: Rubber is a very important industrial crop that is largely produced from the *Hevea brasiliensis* plant. It is widely termed 'natural rubber', because the product rubber can also be produced industrially from petroleum by chemical synthesis. The natural rubber product, however, has some better physical features, including tensile strength and certain unique binding properties that make them suitable and indispensable in the manufacture of certain materials. Tropical countries that produce natural rubber include Malaysia, Indonesia, Thailand, Sri Lanka, India, Liberia, La Cote D'Ivoire (Ivory Coast), and Nigeria, among others.

<u>Coffee:</u> The coffee bean is the product of the coffee plant, *which* exists in several species, including *Coffea arabica*, *Coffea robusta*, and *Coffee liberica*, among others. It is produced widely in the tropics, and

Brazil, Columbia, and several other tropical countries have developed a reputation for the quality and quantity of coffee produced and marketed.

Although coffee is a tropical crop plant, it is consumed in large quantities by temperate countries. Among the largest per capita consumers of coffee are Sweden and the United States, but coffee consumption is so widespread, it is a ritual in the lives of people in almost every country and society. College students are widely known to consume coffee as a stimulant in increasing the number of hours they stay awake and study.

Cocoa: The scientific name of this crop is *Theobroma cocoa*, and it is said to have received this name because of its pleasant taste, causing it to be named "food of the gods." Cocoa is a dicot plant, but it is a peculiar species in that its flower is borne on its main stem, and its fruit, the cocoa pod is bears and grows directly from the stem. The oval fruit with its pedicel bears numerous seeds therein that are the main product of the plant.

The material that surrounds the individual seed is also a tasty material that is consumed locally, but the seeds are only dried and parched before they are consumes.

Internationally, cocoa is the major ingredient in the production of chocholate, a commercial product that largely is a mix of cocoa and milk.

Oil Palm: *Elaeis guineensis*, the oil palm, is a major tropical species. It is mainly grown in the production oil from its fruit, but its fruit is also the main ingredient to a tropical dish that is prepared in a number of ways. The apical meristem is also harvest and used to prepare a vegetable dish in many homes.

The pericarp of the fruit is used to produce the palm oil, and the endosperm of the palm kernel is used to produce the palm kernel oil.

Other Exported Cash Crops: Other exported cash crops include piassava, cotton, jute, and various other products that are exported to a diversity of markets.

Tropical Animals: Production, Marketing and Use

Dominant undomesticated animal species: Hunting is a major activity in rural areas, and animal harvests from the forests meet a major part of the meat requirements in the diet of many folks around the country. The name 'bush meat' is used to refer to harvests from the forests, and we find them in markets, being sold along the highways, and widely consumed in cities and rural communities in West Africa.

The kinds of animals that are sold include deer (or fulentunger), as the popular deer is widely called, wild hogs, monkey, raccoons, birds, porcupines, large rats, and a variety of other meats. The most popular harvest, however, is the deer.

Domesticated farm animals: Cattle, pigs, goats, sheep, and poultry birds such as chickens, ducks, and turkeys are important animals for human nutrition. The need to grow these and other animals for feeding the population cannot be underestimated. Moreover, farming to produce such animals can be a lucrative business.

The commercial animal sector

Cattle: N'Dama cattle are native to Liberia and many West African countries, and they are noted for their resistance to trypanosomiasis, a disease that has limited the number of cattle breeds that would otherwise diversify the production arena and increase beef and even milk production in the tropics. The Brahma cattle from India remain a dominant beef producer, and several breeds are promising milk producers in the tropics.

Poultry: The Single Comb White Leghorn is established as the major commercial poultry breed in many countries in the tropics. Other breeds of note include the Rhode Island Red and the Plymouth Rock, among others.

Goats: Goats are raised everywhere in the tropics, and they contribute significantly to the nutrition of the people. The major advantage of the goat is the fact that it is herbivorous and that it exists with the minimum requirements for care. Goat meat, hide, and milk are obvious products from this animal.

Sheep: Sheep have similar requirements to goats, but they are not as tropical hardy as goats. Sheep also are considered religiously

significant among different populations; therefore, they are not easily commercialized in all of the countries of the tropics.

Other animals: Ducks are among the other animals that are domesticated and raised for meat and eggs, but Guinea fowls and different other poultry-type animals are raised from one section of a country to the other.

Agricultural Production

The rainy season and the dry season are the two annual cycles of the tropical environment, whereas the four seasons of winter, spring, summer and fall typify the temperate environment. In the tropics, diurnal temperatures range from several degrees above 0°C to a maximum around 40°C. For temperate environments, temperatures range from several tens of degrees below freezing (0°C) in the dead of winter to temperatures around 40°C in the heat of summer.

Three (3) problems that are major deterrents to agricultural productivity are:

   a. Environment-Input-Productivity Conflict: The main production season is the rainy season, when the intensity of sunlight is low the quality of sunlight is low to questionable
   b. Light duration is unchanging, making it unfeasible to take full advantage of the scope of adaptation in plants. weeds are plentiful, and herbicides are easily washed off the leaves of plants.

1. Environment and Pest Challenges: Insect pests are very damaging to plant leaves, stems, roots, flowers, and fruits, causing major reductions in crop yields.
2. Harvest and Post-Harvest Losses: Fruits and other food products are easily spoiled, reducing plant yield as well as post harvest losses.

Agricultural Commodity and Product Marketing

The distinction between commodity and product is a significant one that developing countries need to appreciate, because the difference can mean success or failure to agricultural programs and development objectives. Whereas cocoa beans from the cocoa pod are classed as commodities, the chocolate that is produced from the cocoa beans is classed as a product. Products are more pricy than commodities, because of post-harvest processing, and it becomes very important that developing countries recognize this and then exert efforts in developing and marketing various products from the numerous commodities that they produce and sell on the world market.

1.0   World Economics and Africa

Africa is lagging far behind among world economies. Many of the countries of Africa are relegated to the categorization of 'Third World,' and it is important that African countries grow out of this classification. A number of Asian countries that were once classed as Third World countries are now listed in prominent economic rankings. Agricultural development can serve as a major route out of economic underdevelopment, and Liberia and other West African countries can pursue such a route to economic development and success.

To understand where Liberia stands, it is important to compare Liberia with other nations around the world. The

2.0   A Place to Stand
3.0
4.0   Conclusion
5.0

Principles and Practices for the Advancement of Tropical Agriculture and Forestry

TABLE OF CONTENTS

1.0 FOREWORD
2.0 INTRODUCTION
3.0 THE ORIGIN OF TROPICAL SOILS
4.0 TROPICAL SOILS
  4.1 The Nature of Tropical Soils
  4.2 Deficiencies in Tropical Soils
  4.3 Remedies for Deficiencies in Tropical Soils
5.0 TROPICAL PLANT SPECIES
  5.1 Dominant tropical species
  5.2 Primary Food Crop Plants
  5.3 Primary Cash Crops
    5.31 Locally Consumed Cash Crops
      5.311 Sugar cane
      5.312 Peanuts
    5.32 Exported Cash Crops
      5.321 Rubber
      5.322 Coffee
      5.323 Cocoa
      5.324 Oil Palm
      5.325 Other Exported Cash Crops
6.0 TROPICAL ANIMAL SPECIES
  6.321 Dominant undomesticated animal species
  6.322 Domesticated farm animals
  6.323 The commercial animal sector
    6.31 Cattle
    6.32 Poultry
    6.33 Goats
    6.34 Sheep
    6.35 Other animals

7.0 AGRICULTURAL ENGINEERING TECHNOLOGY

Thomas Malthus was an English demographer and political economist who studied population growth and postulated in an essay that was published in 1798 that the world population would outgrow the ability of the earth to produce enough food to feed the population. This dire prediction was to be the cause of c chaos on the planet; however, certain things changed that averted this prediction. Machinery, electricity, and the use of new chemistries in agriculture are major innovations that averted the precautionary calls of Mr. Malthus.

Farm Machinery

Farm tractors are reputed to have first been built by the Minneapolis Steel and Machinery Company (MS&M Co.) that existed between 1902 and 1929. This early establishment of farm power led to the utility of a diversity of power equipment on the farm over the years. Today, there are equipment for plowing, harrowing, planting, transplanting, spreading fertilizer, cultivating, harvesting, milling and drying farm materials and produce.

7.1   Farm Buildings

Residences of the farm family and farm workers are prominent in the planning of a farm, but other buildings including poultry houses, barns for large animals, including goats, sheep and cattle, and swine houses are also important.

7.2   Other Farm Structures

Barns for goats, sheep or cattle, poultry houses, swine houses or pig pens, and even little cages for rabbits, guinea pigs, and other farm animals are important structures.

In addition to buildings, farms have certain kinds of structures that are built to provide additional utility or service. Raised platforms for drying seeds is one type of those utility structures, but they include holding structures for fermenting cocoa, drying or storing cassava, and various irrigation structures to hold tanks, computer systems, utility equipment and various other farm aid devices and equipment.

7.3 Physical Principles and Practices

Farm work can be tedious and difficult, but it is important that farmers and farm workers utilize common sense and their abilities to reason and solve difficult problems or challenges using their knowledge and the power of logic. Principles that guide physical events would be discussed briefly here, then we shall focus on some of the special considerations that farmers make in their daily activities. The farm family and farm workers must always assess their physical strength as well as the tools that are available to them in carrying out their daily duties.

7.4 Solids, Liquids, and Gases

Three states of matter exist. Matter is solid, liquid or gas, and farmers and farm workers must deal with all three of them. When van Helmont (1580-1644) sought the principle of plant growth some 400 years ago, the best recognition of matter was the statement that categorized everything as being of "earth, wind, fire, and water." Today, we know that there are more than 100 elements in the periodic table of chemistry, some being solid, while others are liquid or gas. Combinations of elements form molecules that may be in one state may end up in a different state at various temperatures, pressures, or molecular configurations.

Essential oils such as rose oil, lavender oil, camphor oil, and oils of mint are frequently liquid but easily volatile. Coconut oil, palm oil, palm kernel oil, and even animal fat and tallow exist as solids or liquids at ambient temperatures, dependent upon other factors. Bananas, apples, plums, rice, wheat, and most plant and animal produce are solids, and they are at rest in most situations when they must be dealt with. All of the different states of matter present different challenges in dealing with them, but the farmer and the farm worker are adept in finding solutions to various problems and challenges.

7.41c Gravity

Gravity refers to a force of attraction between two bodies, not regard to magnetic charge, but to their respective masses and the distance between their masses. This is more of the definition that was developed by Sir Isaac Newton in his 1687 publication on the universal law of gravity. For our purposes, the major focus here is the attraction of between the earth and other objects of matter that exist in the realm of earth's gravitational force.

7.41 Simple Machines

In the preparation of the land for farms in the traditional system of agriculture in Liberia and much of West Africa, simple machines were widely used. The axe is a steel wedge that cuts into the trunk to fell the tree. In many cases, scaffoldings that raised the cutter several feet above the ground were typical, and the cutter stood and swung the axe in cutting trees down. The cutlass with wedges that make this utensil sharp is another major implement in the traditional farming system. Nevertheless, these tools are very rudimentary, and more important tools need to be provided for the agricultural system to enable a significant increase in productivity.

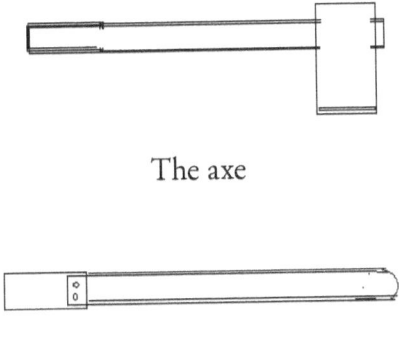

The axe

The cutlass

Modern agricultural equipment

The new agriculture must now begin to utilize the more modern tools, including the power tiller, the rotavator, the planter, and even the modern farm tractor with all of its various attachments and complementary attachments.

7.42   Hydraulic Power: A quick appreciation of hydraulic power is the marvel one witnesses when the lift at a garage raises a car six to seven feet into the air, above the inspecting eye of the mechanic, who moves underneath without a second thought to inspect the undercarriage of the vehicle.

7.43   Sunlight is a tremendous asset for photosynthesis in agriculture, but it is also the power that dries grains and clothes, provides light for the activities of the day, and reduces moisture on leaves and grains to inhibit the attack of bacteria, fungi and other pests from crops and foods.

    7.4   Electricity and Farm Power
        7.51   Hydroelectricity
        7.52   Oil fuel-based Electricity
        7.44   Solar Power
        7.45   Other Power Sources
    8.0   Agricultural Production

Organic Matter, Humus, Tropical Soils and Productivity

Agricultural production is sacrosanct to Liberia development, and it is vital to the development of almost all of the countries in West Africa. A critical factor to successful agricultural production is good soil. Tropical soils can be productive, but the level of organic matter is important to the level of productivity. West African soils are tropical, and organic matter is typically low. Management of these soils for productivity, hence, means that the levels of organic matter must be managed to increase the levels and enhance productivity.

9.0   Agricultural Commodity and Product Marketing
10.0 World Economics and Africa
11.0 A Place to Stand
12.0 Conclusion
13.0
6.0   AGRICULTURAL ENGINEERING TECHNOLOGY
    6.1   Farm Machinery
    6.2   Farm Buildings
    6.3   Other Farm Structures
    6.4   Physical Principles and Practices
        7.46   Solids, Liquids, and Gases
        7.47   Gravity
        7.48   Simple Machines
        7.49   Hydraulic Power
    6.5   Electricity and Farm Power
        7.53   Hydroelectricity
        7.54   Oil fuel-based Electricity
        7.50   Solar Power

The period between October and April covers Christmas and the New Year, and this is the portions of the year that has the greatest amount of solar radiation. The fact is that most days are sunny, and that there is an average daily radiation of at least 5 hours of bright sunshine over this period that is known as the dry season.

The opportunity at present is the significant discovery about solar radiation and the technology that has developed with the science. New solar collectors and storage batteries are sources of enormous electrical and mechanical power for use in agriculture and other industries in Liberia.

7.51   Other Power Sources

Wind power is an older means for the provision of energy, and biofuels are among the newest power sources that are attractive.

References

Chrispeels, Maarten J. and David E. Sadava. 1994. Plants, genes and agriculture. Jones and Bartlett Publishers, Boston.

Lemmens, R. H. M. J., E. A. Omino, C. H. and **Bosch. 2009. Timbers of Tropical Africa – Conclusions and Recommendations – Based on PROTA 7(1): "Timbers." PROTA Foundation, Nairobi, Kenya.**

Miller, Raymond W. and Duane T. Gardiner. 1998. Soils in our Environment. Eighth Edition. Prentice Hall, Upper Saddle River, New Jersey 07458.

Taiz, Lincoln and Eduardo Zeiger. 2002. Plant Physiology. Third Edition. Sinauer Associates, UInc., Publishers, Sunderland, Massachusetts

Salisbury, Frank B. and Cleon W. Ross. 1992. Plant Physiology. Wadsworth Publishing Company, Belmont, California.

Shulze, Willi. 1973. A New Geography of Liberia. Longman Group Limited, London.

www.ingramcontent.com/pod-product-compliance
Lightning Source LLC
Chambersburg PA
CBHW030941240526
45463CB00015B/906